ANOTHER FOURTH POETRY BOOK

compiled by John Foster

Oxford University Press

Oxford University Press, Walton Street, Oxford OX2 6DP

Oxford   New York   Toronto
Delhi   Bombay   Calcutta   Madras   Karachi
Petaling Jaya   Singapore   Hong Kong   Tokyo
Nairobi   Der es Salaam   Cape Town
Melbourne   Auckland

and associated companies in
Berlin   Ibadan

Oxford is a trade mark of Oxford University Press

ISBN 0 19 917125 4 (non net)
ISBN 0 19 917126 2

Books in this series:
A Very First Poetry Book
A First Poetry Book
A Second Poetry Book
A Third Poetry Book
A Fourth Poetry Book
A Fifth Poetry Book
A Scottish Poetry Book
A Second Scottish Poetry Book
Another First Poetry Book
Another Second Poetry Book
Another Third Poetry Book
Another Fourth Poetry Book
Another Fifth Poetry Book

Phototypeset by Tradespools Ltd, Frome, Somerset
Printed in Hong Kong

Contents

The Touch of Sense

Fingers make for miracles
Despite time's warts and hairs.
They smoothe tired children's heads
They ease our daily cares.
They pluck the steely strings
of silkenwood guitars.
They punch the richest meaning
From ivoried piano keys.
Fingers carve new tales
From long-dead stones and trees.
They plant, they reap,
They hold, they wisely heal.
And fingers give long life
To mysteries of words.

John Kitching

Benediction

Thanks to the ear
that someone may hear.

Thanks to seeing
that someone may see.

Thanks to feeling
that someone may feel.

Thanks to touch
that one may be touched.

Thanks to flowering of white moon
and spreading shawl of black night
holding villages and cities together.

James Berry

Black

I shout in praise of black

Ower of nought
The luck of cat
The worth of jet

Velvet's sheen
Hair's shine
Freedom's sash

True sorrow
Soft barrier
Holding back
Some dread tomorrow

Graceful buck
Singing feather
Dancing leather
Wondrous whale
Forest, hill
Flowered and berried thorn

Speed
Grand storm
Cheer of coal
Stately swan

Cowled man
Of holy soul

John Kitching

Through that Door

Through that door
Is a garden with a wall,
The red brick crumbling,
The lupins growing tall,
Where the lawn is like a carpet
Spread for you,
And it's all as tranquil
As you never knew.

Through that door
Is the great ocean-sea
Which heaves and rolls
To eternity,
With its islands and promontories
Waiting for you
To explore and discover
In that vastness of blue.

Through that door
Is your secret room
Where the window lets in
The light of the moon,
With its mysteries and magic
Where you can find
Thrills and excitements
of every kind.

Through that door
Are the mountains and the moors
And the rivers and the forests
Of the great outdoors,
All the plains and the ice-caps
And the lakes as blue as sky
For all those creatures
That walk or swim or fly.

Through that door
Is the city of the mind
Where you can imagine
What you'll find.
You can make of that city
What you want it to,
And if you choose to share it,
Then it could come true.

John Cotton

A Snowing Globe

Inside the glass globe it will snow to order.
A shake and the storm begins,
And while the whiteness swirls about them
The top-hatted gentleman and the full-skirted lady
Are as unmoving as their small brightly painted house
And the wooden tree beside it.
Soon all is still, the tranquil scene restored.
We look in at this miniature enclosed world,
Safe in its predictability,
And give it another shake,
Secure in knowing the storm cannot harm it.

John Cotton

This is My Rock

This is my rock,
And here I run
To steal the secret of the sun;

This is my rock,
And here come I
Before the night has swept the sky;

This is my rock,
This is the place
I meet the evening face to face.

David McCord

Moor-walk

happy-jump
dog lope
skylark winglift
headslicing wind

blow over bee-slope
honeyglow moorside
red-gold landscape
chill heather hill

earthstalking pylons
threaten to conquer
fieldribs shudder
bilberry root clings

honk wild geese
skeining
sweep over daleward
free-tracking
dark smudge
on the sky's pale eye
weeping

Joan Poulson

Notes towards a poem

Sky, grey. Late frost
laces glass, distracts.
Trees, still. Paper, white.

Inside my head
red lava rumbles
in an unquiet earth;
black storm clouds gather,
tigers poise to spring,
a yellowed river
presses at its banks.
Ice binds nothing here
and over lightning flash
and ocean roar
the mountain bursts.
Tigers crash,
white ocean horses
gallop in flood-tide
and down the mountainside
the red-hot torrent pours.

Sky, grey. Late frost.
Trees, laced glass, lost.
Paper, black on white.

Judith Nicholls

To Find a Poem

To find a poem
listen to the wind
whispering words strange and rare
look under stones
there you might find the fossil
shape of an old poem.
They turn up anywhere
in the most unexpected places
look for words that are trapped
in the branches of trees
in the wings of birds
in rockpools by the sea.
And if you find one
handle it carefully
like an injured bird
for a poem can die
or slip through the fingers
like a live eel
and be lost in the stream.
Follow whatever footprints are there
even if no-one else can see them
for clues to lost poems
are waiting to be found
round the next corner
or before you right now.
You may have just missed one
never mind
look again tomorrow
you may find your poem
or your poem
lost somewhere in the dark
may be waiting for you.

Robert Fisher

Full House

Words inhabit my head
like a house. In over-
enthusiasm they tumble
downstairs in a jumble
of arms and legs and
syllables. Some words
shout at me, waving
banners of grammar
from attic windows.

Some I find sitting
prim as sentences
on the settee. Others
hide in cupboards,
come out confused,
complicated as cross-
words. The lazy ones
won't get out of bed
in the mornings, lie

and wait for me to
rouse them thought
by thought. They
ignore me altogether
at times, carry on
as if I wasn't there,
whisper in corners,
upsetting my ideas.

Sometimes words rebel,
won't rock to my rhythms,
move meanings around
like furniture. Mostly
they keep busy polishing
their phrases. At night
they run from room to
room scripting my dreams.

Moira Andrew

18

Ordering Words

Attention   all
            you   words,
GET INTO LINE!
I've had enough of you
Doing what you w
            ill,
STAND STILL!
There are going to be a few changes
Around here.
From now on
You will do
What I want.
THAT   WORD!
You heard,
Stay put.
Youcomeouttoofast
Or per   ulate
      amb
GET IN STRAIGHT!
You are here to serve me.
You are not at ease
To do as you please.
Whenever I attempt to be serious
You make a weak joke.
Always you have to poke
            fun.

AS YOU WERE!
Don't stir.
If ever I try to express
My feelings for someone
You refuse to come out
Or come out all wrong
So sense make none they can of it,
Yet you're so good once they've gone!
Well,
I'm in charge now
And you will say what I tell you to say.
No more cursing
Or sarcasm,
Just state my thoughts clearly
Speak what's on my mind.
Got it?
Right,
F
  A
    L
      L   OUT.

Ray Mather

The Word Party

Loving words clutch crimson roses,
Rude words sniff and pick their noses,
Sly words come dressed up as foxes,
Short words stand on cardboard boxes,
Common words tell jokes and gabble,
Complicated words play Scrabble,
Swear words stamp around and shout,
Hard words stare each other out,
Foreign words look lost and shrug,
Careless words trip on the rug,
Long words slouch with stooping shoulders,
Code words carry secret folders,
Silly words flick rubber bands,
Hyphenated words hold hands,
Strong words show off, bending metal,
Sweet words call each other 'petal',
Small words yawn and suck their thumbs
Till at last the morning comes.
Kind words give out farewell posies . . .

Snap! The dictionary closes.

Richard Edwards

Economy

The words in this book were made entirely from
   recycled letters.
They've all been used before by my elders and
   betters.
Unpack them now, and stack them on your shelf.
Then, when you need them, you can use them
   yourself.

Geoffrey Summerfield

Waste-paper Words

I am the cookie that crumbled;
I'm the pursuit that was hot;
I am the pie that was humbled,
The bolt that was finally shot,
The little game that was rumbled,
The rapidly setting-in rot.

And I am the multitude teeming,
The bull in the china shop,
The spires romantically dreaming
Sweet dreams about a fair cop.
I am the funniness screaming,
I am the pop that is top.

I am the end that is bitter,
The story referred to as tall,
The chat that is married to chitter,
The world that we all know is small.
I'm the words they scatter like litter
And I have no meaning at all.

Vernon Scannell

Ronald/Donald

Ronald Derds (or was it Donald Rerds?)
Was a boy who always wixed up his merds.
If anyone asked him; 'What's the time?'
He'd look at his watch and say, 'Norter past quine.'

He's spoken like that ever since he was two.
His parents at first didn't know what to do.
In order to understand what he'd said,
His father would get him to stand on his head.

But this didn't work, something had to be done,
So Ma and Pa Derds learnt to speak like their son.

'Mood gorning,' he'd cry, as he chat in his sair.
'Gorning,' they'd answer, without hurning a tair.
And Ron's Mum would say, 'Get a nice brofe of led,'
For Ron to return with a loaf of fresh bread.

Then one special day, young Ronald's voice broke.
He found it affected the way that he spoke.
'Good morning,' he said as he sat in his chair.
'Gorning,' said the others and started to stare.

From that moment on, things just got worse.
The harder they tried, they just couldn't converse.

Ron said to his parents, after a week,
'It's driving me mad, the way that you speak.
I can't understand a word that you say.
You leave me no option, I'm leaving to-day.'

So Ron joined the Navy and sailed to the Barents,
To get as far away as he could from his parents.
And although this story all seems rather sad,
Ron occasionally visits his Dum and his Mad.

Rod Hull

New Leaf

This is the first day of my new book.
I've written the date
and underlined it
in red felt tip,
with a ruler.
I'm going to be different
with this book.

With this book
I'm going to be good.
With this book
I'm always going to do the date like that,
dead neat
with a ruler
just like Christine Robinson.

With this book
I'll be as clever as Graham Holden
get all my sums right, be as
neat as Mark Veitch;
I'll keep my pens and pencils
in a pencil case
and never have to borrow again.

With this book
I'm going to work hard,
not talk, be different
with this book
not yell out, mess about
be silly
with this book.

With this book
I'll be grown-up, sensible
and everyone will want me,
I'll be picked out first
like Iain Cartwright:
no-one will ever laugh at me again.
Everything will be

different
with this book. . . .

Mick Gowar

Eyes

I wish that when
I was at school
That I'd been able to see
That I was as clever
As the clever ones were
And that they were as stupid as me.

Julie Holder

The New Teacher

What big eyes you've got, Teacher!

A new design, spring-swivelled
To see round corners in cloakrooms.
There's another pair in the back of my head.

You have very big ears.

Twin lie-detectors, these,
To filter the excuses of truants,
Lazy workers and late-comers.

And your nose, does it light up?

It will flash if anyone
Except the caretaker smokes a cigarette
Behind the bicycle sheds.

Why do you need such big teeth?

I am programmed to eat crisps,
Sweets, apples, birthday cake, seaside rock,
School dinners and sometimes children.

In my tin skull is a hundred years
Of homework to be done
In neat handwriting, with no mistakes.

Will you be taking 2G?

Irene Rawnsley

Silent Shout

Don't want to do
As I am told
Don't want to start my brain
From cold
I feel big and bad and bold
Let me out.

Don't want to work
Don't want to play
Don't want to be here anyway
I won't stay
I'll run away.

Don't want to be
A meek and mild
Don't want to be
Where my name is constantly dialled
I'll run wild.

I sit still through the others' clatter
I scowl and bite my thumb.
When a teacher asks me
'What's the matter?'
I am dumb.

Julie Holden

Classroom Globe

We strung our globe from the rafters
then watched as the continents span.
We were dizzy with faraway places,
they swam before our eyes.
Everyone wanted to take a swipe at
the planet, to roll the world, to cause
global chaos. We laughed at the notion
of some great hand, sweeping down
avalanches, rolling earthquakes round
Africa, knocking elephants off their feet.
Then reasons were found for leaving seats,
to touch, or tilt, or hit heads on the planet,
squaring up to the world like March hares.
We talked of how the Earth had been damaged,
leaving it bruised, sore from neglect,
and Jenny, who feels sorry for anyone and
anything, lent her brow against the planet,
and felt the sorrow and pain of Earth
in a cold hard globe.

Brian Moses

Request

I pray you, teacher,
Teach me so
That I will always
Want to know.

John Kitching

Cardinal Ideograms

0   A mouth. Can blow or breathe,
be funnel, or Hello.

1   A grass blade or a cut.

2   A question seated. And a proud
bird's neck.

3   Shallow mitten for two-fingered hand.

4   Three-cornered hut
on one stilt. Sometimes built
so the roof gapes.

5   A policeman. Polite.
Wearing visored cap.

6   O unrolling,
tape of ambiguous length
on which is written the mystery
of everything curly.

7   A step,
detached from its stair.

8   The universe in diagram:
A cosmic hourglass.
(Note enigmatic shape,
absence of any valve of origin,
how end overlaps beginning.)
Unknotted like a shoelace
and whipped back and forth,
can serve as a model of time.

9   Lorgnette for the right eye.
In England or if you are Alice
the stem is on the left.

10   A grass blade or a cut
companioned by a mouth.
Open? Open. Shut? Shut.

May Swenson

31

A Boy Wrote a Poem

A boy wrote a poem,
It was for homework from class,
He wrote about cliff-tops,
And how the winds pass.
He just let it flow
From his head to his pen,
But his spelling was bad,
'C, do this again!'

A boy wrote a poem,
And thought of his mark.
And this time he checked it
And wrote of the dark.
He changed and corrected,
Gave it in the next day,
He got 'B + Good effort'
And threw it away.

When he wrote of the oceans,
They gave him an 'E'.
They gave him an 'E'
For the tides in the sea,
'What does this mean?'
Said the boy to his work,
'Does it mean I'm just lazy,
Does it mean I'm a berk?'

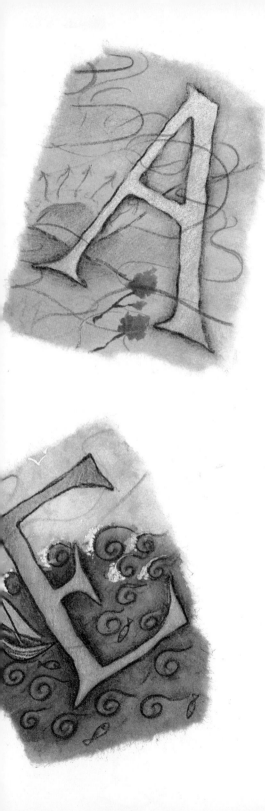

When he wrote about sunrise,
They gave him an 'A'.
They gave him an 'A'
For the dawn of the day,
'What does this mean?'
Said the boy to his paper,
'Am I meant to be happy,
To leap up, cut a caper?'

'What is this letter?
It is nothing to me!
It doesn't mention the good bit
At the end of verse three.'
And so thought the boy,
But he couldn't be sure,
So he looked at his shoes
And the tiles on the floor.

Then, before long,
At a certain time,
They asked for the marks
Of the efforts in rhyme.
To be written down
In a large orange book,
In symmetrical lines,
To be read at one look.

And he sat at the back,
At the back of the room,
Among the new novels,
The display work, the gloom.
And they asked him his marks,
And he read them ashamed,
When he got to his worst
'Tut tut,' they exclaimed.

Oh, he thought.
But I don't see what those marks have
Got to do with my work.

He still doesn't understand.

Nicholas Chapman

Acrobats

Most of the numbers look
As if they could never move
Like 1 that stands straight up
To start the pages in a book.

2 is on its knees,
4 sits down to wait
And 3 is looking for 5
To turn it into 8.

7 raises an arm to wave
When you turn its page
And 8 looks far too fat
To ever misbehave.

But sometimes I imagine
While 7 and 8 stood still
6 turned a somersault
And ended up as 9.

Stanley Cook

The Pond

There was this pond in the village
and little boys, he heard till he was sick,
were not allowed too near.
Unfathomable pool, they said,
that swallowed men and animals just so;
and in its depths, old people said,
swam galliwasps and nameless horrors;
bright boys kept away.

Though drawn so hard by prohibitions,
the small boy, fixed in fear, kept off;
till one wet summer, grass growing lush,
paths muddy, slippery, he found himself
there, at the fabled edge.

The brooding pond was dark.
Sudden, escaping cloud, the sun
came bright; and, shimmering in guilt,
he saw his own face peering from the pool.

Mervyn Morris

Ice on the Round Pond

This was a dog's day, when the land
Lay black and white as a Dalmatian
And kite chased terrier kite
In a Kerry Blue sky.

This was a boy's day, when the wind
Cuts tracks in the sky on skates
And noon leaned over like a snowman
To melt in the sun.

This was a poet's day, when the mind
Lay paper-white in a winter's peace
And watched the printed bird-tracks
Turn into words.

Paul Dehn

Miller's End

When we moved to Miller's End,
   Every afternoon at four
A thin shadow of a shade
   Quavered through the garden-door.

Dressed in black from top to toe
   And a veil about her head
To us all it seemed as though
   She came walking from the dead.

With a basket on her arm
   Through the hedge-gap she would pass,
Never a mark that we could spy
   On the flagstones or the grass.

When we told the garden-boy
   How we saw the phantom glide,
With a grin his face was bright
   As the pool he stood beside.

'That's no ghost-walk,' Billy said,
   'Nor a ghost you fear to stop –
Only old Miss Wickerby
   On a short cut to the shop.'

So next day we lay in wait,
   Passed a civil time of day,
Said how pleased we were she came
   Daily down our garden-way.

Suddenly her cheek it paled,
   Turned, as quick, from ice to flame.
'Tell me,' said Miss Wickerby.
   'Who spoke of me, and my name?'

'Bill the garden-boy.'
      She sighed,
   Said, 'Of course, you could not know
How he drowned – that very pool –
   A frozen winter – long ago.'

Charles Causley

Our Farm, January

it was cold enough to freeze a snowman's legs off
it was colder than I'd ever known before
and one morning when I got up it had happened
summat as I'd waited twelve month for

the pewtered sky had been relieved its burden
the frozen grip let go its hold at last
and our farmyard, down the daleside
all the valley out to Thirsk
was eiderdowned and safe-tucked-in
soft and white and hushed

I knew I'd best keep quiet about my feelings
it i'n't easy looking after stock in this
and it's hard enough in Summer to make a go up here
i' Winter it can crack your heart, reduce a man to tears

but I love to see it look like this
our dale – and out beyond
untouched, and pure, and all snugged-down
feathered o'er i' snow

Joan Poulson

Invaders

Today it is snowing:
The starlings are out in force,
Bullying the sparrows,
Their bayonet beaks
Commandeering the breadcrumbs.
Like stormtroopers
They take over the garden,
Asserting
That might means right.

John Foster

Snow

Touch me, I am more soft
    than wool.
        I fall
multiplying – but make
    no sound
        at all.

I knew this earth before
man came, and lay
unique and nebulous
from bay to bay;

No footprint scarred my smooth
complexity
in that stern, arctic night.
There yet may be

Years when your motorways
    once more
        will lie
snow-silent, giving me
    back
        to the sky.

Jean Kenward

Thaw

You cannot describe it
labour how you may –
that soft, intrinsic sighing
as if there were someone near –
is it the snow out there?
Snow crystals, dying?

You will find no-one
Pan-like in the woods
whose footmarks pock your way,
though you might follow that
deep-stippled, darkening spoor
till prime of day;

Only those sinister
movements among the slick
and moisture-dropping trees
tell of a presence that
heeds neither exits here
nor entrances.

Jean Kenward

Sea Spring

In Spring
　　the willow
is captain.
　　　Rigged in silver,
commanding
　　the wave-dashed
banking.

Around and above,
　　tall trees sway,
mast-like, pennanted
　　by the ragged form
of a crow. Lashed,
　　held there in
the squally blow.

Joan Poulson

Seven Days of Spring

First, doors are opened onto light
        filling the vast halls of the hills
        until they tremble.

Second is a new wing of air
        its daring livewire stunts begin
        the grand ovation.

Third is water, gathered into choirs of streams
        running through clearings
        with the loud elated voice of larks.

Fourth, earth lays out clothes, its summer best
        green telegrams appear saying:
        'we accept. are on our way'.

Fifth and the broadcast now is live
        an audience blossoming with oohs and ahs
        sees fields make welcome every guest.

Sixth has people turning in the streets
        to the first stranger they see
        to say how marvellous it is.

The seventh eases in
        and now at last the moon appears, over rooftops,
        woodslopes strewn with anemonies,
        moving slowly, bridal white.

John Sewell

The Old Canal

On the stillest of still days
We lay on the towpath to look
Down into clear depths.
We saw ourselves looking up, drowned.
We tapped the water with finger-tips
And our faces wobbled like jelly.
Tapped again and they broke up.

Under the bridge, cool and dark:
Light bounced off deep water
To play on the bridging vault
Of curving brick: a ghostly
Play of wavy trembling light,
Oscillating, oscillating.

Every day, the dare:
To cross the bridge, outside the wall.
A bare toe-hold on the lip of bricks.
Clung to the parapet, arms outstretched,
Fingers aching, scratching a hold.
Reached safety, trembling.

Near factories, canal-water
Always a rusty red soup,
Stinking of bitter iron.
We threw sticks: 'Fetch it, boy!'
He did, and stank to high heaven
For at least a week.
'Stay outside, pongy dog!'

Old sacks, stacked with bricks,
Took dead dogs to the bottom,
Lay in deep mud, disintegrated.
The corpses rose, ballooned,
The colours of death.

In clear stretches, stickle-
Backs, big-eyed, flickery.
We netted them, scraped off
The limpets, little plates
Like cancers: gristly
Grizzled and grisly.

Water-boatmen, long-legged,

Balanced on dimples of water,
Performing their ordinary miracle.

Dragon-flies, encased in enamel,
Decorated the flags with their pennants.

Beetles, shoe-black shiny,
Shot up like rockets,
Surfaced, gulped air.
Then down again
Like suicidal dive-bombers.

Best of all, the barge-horses,
Thudding, treading
For purchase, straining
Shouldering, muscles at a stretch.

Ropes of grassy saliva
Hung, swinging.
And the barge, dark as night,
Looming like a giant coffin.
Always at the stern, silent,
A man or woman at the tiller,
Dark and silent as an undertaker.

Worst of all, my brother getting lost.
We searched the towpaths, bridges.
Was he already under water?
At ten, he couldn't swim,
Was crazy about canals.

At nightfall, he was still not found.
Fears bubbled up
Inside our minds,
Like the frantic gurgling
Of a drowning boy.

His face was in my mind's eye,
White, silent, staring from eyes
Now blind, wide open, dead.

When he came strolling home,
A torrent of questions.
'What's all the fuss about?'
He asked. He'd found
A new exciting friend
With an even more exciting
Even newer model railway.
Shunting, moving signals,
Changing points,
They'd lost all track of time.

He was sent upstairs.
Then thrashed.
There was a sudden rush
Of sheer vexation,
The sharp anger of relief.

For months, every night,
Falling asleep, I saw
His drowning face.

Geoffrey Summerfield

Carps

There's carps in Boxer's Lake – they said,
With scales as gold as crowns,
With leathered lips,
And teeth like chips,
And gills as red as blood.
There's carps in Boxer's Lake – they say,
The big boys told us so –
they're special ones
with eyes like plums
and tongues as long as lies.
They live deep in the middle,
They live in deepest mud,
And suck down passing cygnets. . .
and suck their cygnet blood.
They've been there for a thousand years
A snorting bubbling brew
who snap the anglers' fibre rods
and eat the ducklings too.
So–
Maurice went to catch one.
I went with him one night,
Along with Kenny Murrell
and Kenny Murrell's bike.
. . .

A triple hook with wormcake bait,
Line as strong as wire
and six lead weights
and extra baits
a float as red as fire.
. . .

Four hours we stood in Boxer's Lake
Four hours our wellies leaked
and let in nasty squirmy things
that looked like butcher's meat.
We watched our fireglow balsa cork
we watched it still and quiet . . . a signaller of armies
on a silent summer's night.
The cygnets settled on their nests
they crouched beneath the wing . . .
the evening bat came out to play

the blackbird ceased to sing.
Our wellies filled another inch,
The reedbed creaked its song
and Maurice in his husky voice
huskied :
            'Won't be long – '
The crimson float it turned to black,
The water ceased to shine,
and Kenny – in his Kenny way
began his Kenny whine.
There's nothing here.
There never is . . .
    I've never caught a thing
you won't catch much with fishing-line
that looks as thick as string.
So Kenny cycled home you see
but me and Maurice stayed
and watched the float,
and changed the bait,
and watched the shadows go.
That night time lake filled up with noise,
We heard the sucking carp,
and that is when I had to go,
before it got too dark.

THERE'S
CARPS IN BO[...]
LAKE[...]
WITH S[...]
GOLD AS CR[...]
LEATHER[...]
AND TEE[...]
AND GILL[...]
[...]ED AS BLOOD
[...]RPS IN BO[...]
SAY ~ THE[...]
[...]THEY'RE[...]
[...]H EYES
AND TONGUES
THEY LIVE IN
[...]UCK DOWN PASSING
[...]CK THEIR
[...] BLOOD

SAY
AS
WITH
LIPS
CHIPS
AS
THERE'S
LAKE
BIG BOYS
SPECIAL
LIKE PLUMS
AS LIES
AS LONG
DEEPEST   MUD AND
CYGNETS ...AND

But Maurice stayed – to catch that fish,
His Mum was never in,
and didn't care about the time,
or where her boy had 'bin.
. . .

So Maurice stayed, and caught that carp.
He caught it dead of night.
A thousand yards of tangled line.
Two hours of frightful fight.
He told me all about that carp,
in morning prayers – next day
and how he'd got this massive bite
near where the cygnets lay.
He told us of its stinking mouth,
Its lips adorned with hooks,
its fleshy tail
its evil eye
and fungoid festered looks.
But:
I never saw that massive carp
or saw his cork float fly
or saw its withered lumpy lips
or saw its evil eyes.
. . .

He wrote it up in newstime,
He wrote of tangled lines –
and splashing mud,
and carpfish blood,
and carpfish howls and whines.
>He wrote of shriekings in the lake,
>He wrote of carpfish eyes,
>And Miss Gardner in her teacher way
>red biroed
>See me!
>LIES.

Peter Dixon

Duppy Story Time

De breeze blow cold, de night it dark,
But none a we no worry,
For we siddung roun de fire
An Granny a tell we duppy story.

'Me tell unoo how Mass Joseph bull-cow
Tun eena rolling calf?
Im lick dung Mass Joseph flat a grung
An den dis start to laugh.

An den de sinting yeye dem
Flash fire like de debil own,
An po' Mass Joseph couldn' move
Dis lie dung pon de dirt a moan.

A Parson Walters save 'im,
'Im ben dissa pass, me dear,
'Im dis rebuke de rolling calf
An de sinting disappear.

Until teday dem no fine de bull-cow,
An Mass Joseph still cyan talk,
An im dis a stagger bout de place
Like baby dis a larn fe walk.

Amy did fine one bottle,
Green, wid a funny shape,
She notice sey it cork tight
An fasten up wid tape.

She pull i', den notice one funny smell
Like dead flesh, rotten bad,
She feel sinting box har roun har head,
From dat day de pickney mad.

It noh good fe walk bout a night time
For nuff duppy deh 'bout,
Dem look like people, tell yuh "how-de-do",
But ef yuh opin yuh mout

Dem drag yuh gawn a cemetry,
An yuh cyan eben he'p yuhse'f,
Before mawning light an cock start crow
None a yuh noh lef.

54

Yuh fren an family know someting wrang
But dem noh know a hat,
Dem wonder how yuh jus' disappear,
Dem try fe fine – A wha dat?!

Come back pickney! Whey unoo run gawn?
A how unoo all fool-fool soh?
Unoo mean unoo a go kill unoose 'f
Ova one deggeh patoo?'

Marie, quick turn up de lamp,
Meck sure yuh lock de door,
Now help me look roun fe me heart
Me think it jump out pon de floor.

Valerie Bloom

55

The Magic Wood

The wood is full of shining eyes,
The wood is full of creeping feet,
The wood is full of tiny cries:
You must not go to the wood at night!

I met a man with eyes of glass
And a finger as curled as the wriggling worm,
And hair all red with rotting leaves,
And a stick that hissed like a summer snake.

The wood is full of shining eyes,
The wood is full of creeping feet,
The wood is full of tiny cries:
You must not go to the wood at night!

YOU MUST NOT GO TO THE WOOD AT NIGHT!

He sang me a song in backwards words,
And drew me a dragon in the air.
I saw his teeth through the back of his head,
And a rat's eyes winking from his hair.

The wood is full of shining eyes,
The wood is full of creeping feet,
The wood is full of tiny cries:
You must not go to the wood at night!

He made me a penny out of a stone,
And showed me the way to catch a lark
With a straw and a nut and a whispered word
And a pen'north of ginger wrapped up in a leaf.

The wood is full of shining eyes,
The wood is full of creeping feet,
The wood is full of tiny cries:
You must not go to the wood at night!

He asked me my name, and where I lived;
I told him a name from my Book of Tales;
He asked me to come with him into the wood
And dance with the Kings from under the hills.

The wood is full of shining eyes,
The wood is full of creeping feet,
The wood is full of tiny cries:
You must not go to the wood at night!

But I saw that his eyes were turning to fire;
I watched the nails grow on his wriggling hand;
And I said my prayers, all out in a rush,
And found myself safe on my father's land.

Oh, the wood is full of shining eyes,
The wood is full of creeping feet,
The wood is full of tiny cries:
You must not go to the wood at night!

Henry Treece

In the Woods

In the woods we would quietly skirt
That area where we could just glimpse,
Through the brambles,
A makeshift shelter of tarpaulin and sticks,
And, occasionally, movement or a drift of smoke
We never went nearer,
Nor thought of doing so.
It wasn't fear so much,
Though that was part of it,
But the sense of a secret, alien life
Which we furnished with our imagination,
And was worth keeping a mystery.

John Cotton

Goodnight Stephen

At first it was the smell,
the smell of a torch
drifting up like mist through the field.

Then it was the sound,
the sound of a torch,
a noise like a torchbeam unzipping the tent.

Now it's the weight,
the weight of a torchbeam
across the sleeping-bag onto my face.

I must be asleep
but I think I'm waking up.

The stink of the torchbeam
smells awful, smells scary.
A torchbeam feels spiky.

The taste of the torchbeam
tastes rotten, tastes fishy.
A torchbeam feels chilly.

I think I'm waking up.
I can't be asleep.

Then it was the smell,
the smell of a pipe
through the flap of the tent,

and it was the sound,
the sound of my dad
saying 'Just checking. Goodnight.'

Now it's the weight,
the weight of my head
on the pillow as darkness returns.

Ian McMillan

Tell me about your dream

'It's always the same you see,
Never varies, and always
Leaves me sweating with fright.
Talk you through it?
Well, I'll try.
Yes, that's right, I'm at a football match,
And it's a big game, really big.
We're all waiting,
Waiting for the teams to emerge.
Then suddenly over the speakers
This voice comes, very clear,
And everybody goes quiet.
Do I remember what it says?
Well yes; two of the players
Have missed the train
So they're a man short, it says,
And if anyone
Happens to have his boots
They'd be grateful if he'll play.
Course I have,
Always bring my boots
Because I know from my dream
This could happen.
What? That's right,
I don't know if it's a dream or not.

In a flash I'm over the fence
And suddenly I'm in the dressing room
Surrounded by famous players
All shaking my hand
And clapping me on the back.
As they run out
The manager holds me back.
He looks me in the eye and says,
'I've kept you back
To appoint you captain.
I want you out last
To enjoy alone your moment of glory.'
He hands me a number nine shirt.
'Wear it with pride lad';
And then I'm off
Up the dark tunnel
To my moment of glory.
My studs rattle on the concrete
And the crowd's roar
Is like a great beast
Breathing far away.
I run down a white corridor
Now they're calling my name
Soon I'll trot out into the sun.
Suddenly I come to a wall.

I don't know whether to go left or right.
I take the right one
But the right one
Turns out to be the wrong one.
It leads to another white corridor
And another white corridor
And another white corridor;
The crowds far away now
And I'm hopelessly lost
And suddenly I wake up
Shaking with fright.
That's it.
What have I done about it?
Well, I don't take my boots
To matches any more because,
Well, you never know.
Has it helped to talk about it?
Yes, I suppose it has.
Do I want to ask any questions?
Well, there is one thing.
I don't know if you can tell me:
Is this a dream?

Gareth Owen

Marathon Man

The thin-legged man
Totters through the suburbs
With the other escaped convicts
Past chemist shops and parked cars
It does not seem to occur
To the passers-by
To inform the authorities
Of the mass break-out
And some even hand them drinks
Like all the other prisoners
The thin-legged man
Has a number on his back
So that the police can identify him
But even though he is white-faced
With unutterable weariness
The pursuing cars
Seem incapable of catching him.

Gareth Owen

High Jumper

The high jumper
Incessantly treads
Small grapes beneath his heels
Cools his wrists
In currents of air
And bounces bounces
In an arc towards the bar
Touching the hot coals
Beneath his feet
As little as possible.

The Long Jump

The long jumper
Sucks in draughts of air
Through an unseen straw
Shakes the water from his hands
And sways sways
In a gale only he can feel
Leaping off the board
He bicycles on air
Before the quick sand
Draws him down.

Gareth Owen

64

The Hammer Thrower

That is the enemy there
Identifiable by their blue blazers
My job is to kill their leader
With this hammer on a chain
Round and round I spin
Grinding my heel
And puffing up my anger
With each turn
When my rage is at its highest
I curse my tormentors
With the war cry of my tribe
And launch the hammer at my enemies.
Lesser men
They cunningly stand just out of range
Then scurry to the place
Where my missile explodes
Out of pure spite
They measure my failure
And broadcast it
To the world.

Gareth Owen

The Relay

This is the scroll
Containing the important message
Each of us runs
'Till we can run no more
Then at last gasp
We pass it on
Eventually the message
Comes full circle to the sender
When he reads it aloud
We laugh and dance
At the news of our victory.

Gareth Owen

Cross-country

On and on
Through the snow we run
Puffing and panting
It's not much fun
Over the fence
And into the stream
Mouths obscured by commas of steam
Plimsolls soaked
Shorts splattered with mud
Freezing to death
'For our own good'!
Plodding along
Mile after mile
Under the gate over the stile
Legs like jelly
Feet like lead
Wet hair slapping
Against the forehead
Defying the cold
For just under an hour
Before burning to death
In a scalding hot shower!

Ray Mather

Boxer man in-a skippin' workout

Skip on big man, steady steady.
Giant, skip-dance easy easy!
Braad and tall a-work shaped limbs,
a-move sleek self wid style well trimmed.
Gi ryddm yu ease of being strang.
Movement is a meanin and a song.
    Tek yu lickle trips in yu skips, man.
    Be dat dancer-runner man.

Yu so easy easy. Go-on na big man!
Fighta man is a ryddm man
full of de go, free free.
Movement is a dream and a spree.
Yu slow down, yu go faas.
Sweat come oil yu body like race horse.
    Tek yu lickle trips in yu skips, man.
    Be dat dancer-runner man – big man!

James Berry

Let you body go

Me say get you bones movin, let you body go.
Get you heart beating, speed up no go slow.

    Stick out you toes
    feel you blood flows

    Stretch you lickle leg
    hang it on a peg – aya

    Bend you in the middle
    it no such a fiddle – bim

    Straighten out you back
    Fe no saggin like a sack

    Twist you floppy neck – gribbit
    say man what the heck

    Stickle out you tongue
    Then you song is sung

Me say get you bones movin, let you body go.
Get you heart beating, speed up no go slow.

John Rice

Break/Dance

I'm going to break/dance
turn rippling glass
stretch my muscles
to the bass

Whoo!

I'm going to break/dance
I'm going to rip it
and jerk it
and take it apart

I'm going to chop it
and move it
and groove it

Ooooooh I'm going to ooze it
electric boogaloo
electric boogaloo
across your floor

I'm going to break/dance
watch my ass
take the shine
off your laugh

Whoo!

I'm going to dip it
and spin it
let my spine twist it
Ooooh I'm going to shift it
and stride it
let my mind glide it

Then I'm going to ease it
ease it
and bring it all home
all home
      believing in the beat
      believing in the beat
      of myself

Grace Nichols

Flight Patterns

John Sewell

Bird-cloud

poppy-head explosion
uplifts
chattering
from the hedge

settles
into the verge
ahead

still scolding
still outraged
at my disturbance

Joan Poulson

The Starlings

The sparrows come in a chattering crowd,
workaday chaps in their dun overalls
they peck fast, first come, first served,
the rule they follow. Suddenly

they scatter. The starlings, in a shine,
in a scintillating rush, descend,
jewelled in the sun. Worldly-wise,
they waited to see that all was safe

before they claimed what they knew was theirs.
Swaggering rogues, as bright of eye
as of feather, shrill-voiced, sharp-beaked,
they squabble greedily. Full of Dutch

courage, they are still alert – they fear
my old tom, he's their sworn enemy –
they clear up every last crumb, wait,
strutting about, chortling, whistling,

churring, using their many voices
to defy him, pretending to forget him,
they're much too busy discussing business
than to bother with such a bedraggled

old ragbag. He rushes in a snarling rage
from behind the bushes. They stand,
fly up in his face, perch in the lilac,
mew like a cat the more to madden him.

Albert Rowe

Dawnsight

Skimming
its
lost eye

a sandpiper
brushes
the tarn awake.

Fringing
its
gold iris

lashes of reeds
salute
the sun.

Walking Out

Walking out
to taste
the morning

a sky
harebell
delicate

dew
on the
tongue's leaf

beyond
the world's ear
a song.

Geoffrey Holloway

Blue Tits

have lemony waistcoats, clown-white, round-bearded faces,
skullcaps and eyes that glint like ball bearings

somersault, trapeze, trampoline – twig to leaf, leaf to twig,
(evanescent, light as raindrops)

clamp under coconut caves, straddle cake smithereens,
their tiny rakish claws steel-delicate –

throw eyes over shoulders, flick back in time
to snap scissor-beaked at scrounging relatives

but themselves only snatch between eyeshots –
not to be part of tomorrow's feline gut.

Geoffrey Holloway

The Whole Duck

A duck's head under water
Is deep in thought
Which makes its body shorter
As thinking ought

Until, quick as a flash,
The whole duck reappears
And with a little splash
Floats its ideas.

John Mole

Nature Study

This butterfly
  we couldn't identify
    pitched
  its bright tent
on a roadside flower.

For a full minute
  outstretched wings
    bloodied
  the morning air
in studied symmetry.

Our eyes ached
  with raw colour
    remembered
  pattern and shape
against eventual flight.

It drifted away
  and precise geometry
    lingered
  like an after-image
in the yellow heat.

Moira Andrew

79

Palmtrees

a long time ago
they grew to love the sun
so much they simply
stood and dreamed
until their claws
turned roots and they could
no longer fly

and then small mammals
learnt to climb
up into their crutches
to steal their eggs
before they laid them

great flocks of them
flutter by the shore
they do not notice
the small mammals

the sun shines on them and
they are still dreaming

. . .

in the enormous
room of the dusky plain
worn by their efforts
against the cobwebs
of dust and haze, like
tattered feather dusters
the palmtrees are propped
up against an horizon
glowing red raw with
the ageless domed lamps
of cane fires

Dave Calder

Oranges

which came first
the colour or the fruit

• • •

through the archway
a tarnished moon.
in the wicker basket
green oranges huddle
in unweeping melancholy

• • •

in the long grass a ripe
orange. its heart
secretly stolen by ants

• • •

in the tree a young
boy and
the oranges. both
will come down
together

• • •

when they are ready to be picked
the oranges
stop pretending to be leaves

• • •

the orange on the table
drew all the light in the room into it
and still it did not shine

• • •

even in the hand the orange
maintains an air of
resolute inviolability

• • •

her fingers pressed just so hard
into the orange

flesh into flesh
her mind was elsewhere

• • •

the torn skin
shards
of a broken pot

• • •

nothing is shared
as simply as
an orange

Dave Calder

81

Villanelle

This year I left the apples on the tree
for birds to pick at for their autumn food,
not even gathering one of them for me.

Watching them probe and balance I could see
that I had done the very thing I should.
This year I left the apples on the tree,

took none for keeping, but birds feeding free
seemed to be something sensible and good.
Not even gathering one of them for me

from all the scarlet crop that brilliantly
shone against yellow leaves and lichened wood
this year I left the apples on the tree.

December: still I watch the two or three
left red as baubles for a Christmas mood,
not even gathering one of them for me.

For once, it's true, I could more easily
forgo them; they were small, not very good
this year. I left the apples on the tree
not even gathering one of them for me.

Pamela Gillilan

The Watertower

My father went sticking or blackberrying
Or gathering sweet chestnuts from the wild orchard
Of hedgerows and woods, like someone exercising
An ancient right on common land,
I with my rabbit-food sack, blotched brown
With dandelion milk, alongside.
Picking our way at the edge of a field of oats
Wasting away like that in the parable
That fell on stony ground or going through
The ruins of some remote abandoned cottage
Or over the waste of an overgrown quarry,
He related at length how it happened,
The history of local people's lives
That would never get written down.
From a rise we could see the watertower,
Stilted on concrete above the plain,
Whose building he once had charge of,
Wild daffodils he brought from the site
Still flourishing in our flower garden.
We looked at it without a word, reminder
Of better days before he was out of work.

Stanley Cook

Growing Pains

The twelfth of August.
The sun-baked ground brown concrete,
the yellow grass sparse hair
on an old woman's head,

ants like freckles
twitch between the thin wisps:
everybody lies and gasps
– except my father.

Swollen with energy he capers,
bulgy over crimson boxer shorts;
I cringe behind the apple tree –
how stupid he looks.

How childish he is:
How could he:
what if someone sees him
dressed like that –

gamboling like a loon
all flab and chest hair,
long short grey socks
and rubber flip-flops?

Then Dad lifts the rabbit from its run
and trots around the garden,
the big black buck held firm
against his naked chest,

giving it a guided tour – 'These are
the cabbages – yum, yum – and there, the goozygogs. . .'
My Mum, flat out beneath a William Pear
sees danger: 'Do be careful, dear.

If Sooty jumps – you know how strong
his back legs are.' My Dad
stares into the rabbit's eyes:
'He wouldn't hurt his Daddy,
                                        would he?'

The rabbit, bored of being Bunny,
leaps free. Kicking back
his claws leave four great
bleeding weals across my father's chest.

My mother crows – 'I told you so' –
and leads the way in doors to
bathe the wounds in antiseptic.
Me – so much more adult,

embarrassed, bored and cross
beneath the tree –
'Serves him right, serves him right,
serves him right!'

From somewhere in the house
I hear a bellow
as the antiseptic stings:
'Big Baby!'

Mick Gowar

85

The Invalid

My mother was ill for years
and years. It started in a small way
when I was twelve, and stayed home
for several days to look after her.

From that time on whenever
I was late for school, sometimes
even too late for Assembly,
I'd arrive, not guiltily,

but with a sad and careworn face.
The teacher would ask how
my dear mother was that day.
My reply was seldom optimistic.

I'm happy to say my mother enjoys
life to the full. She doesn't know
that she was a famous invalid, until
that sudden recovery the day I left school.

Pamela Gillilan

Grandfather Gavin

my grandfather gavin
kept his morris minor
in a wooden boathouse
miles from the seashore
but he drove it like
you'd steer a boat
it bounced, bobbed and bellied
only just afloat.
and up the rolling waves of hills
and down the other side
we sailed, as fast and thrilling
as a rollercoaster ride.

my grandfather gavin
was known near and far.
i think people stayed indoors
when he drove his car.
my grandfather gavin
had a round bald head
it was rounder and shinier
than his morris minor
and he parked it in his bed.

Dave Calder

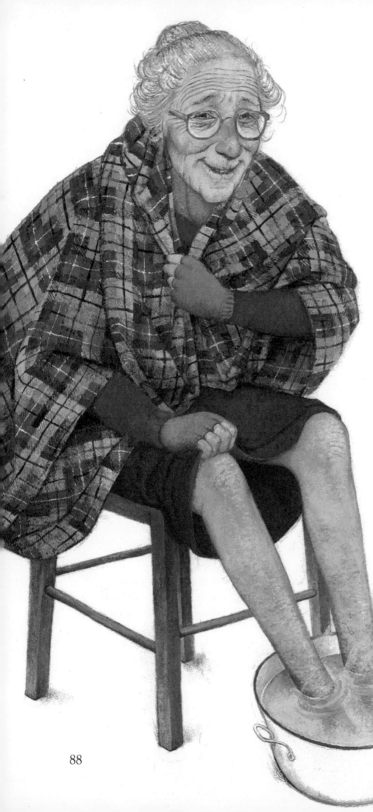

Gran

In winter gran got chilblains.
Calling to see her
On the way home from school
I'd catch her with her skirt
Up round her knees,
Her feet immersed
In a white enamel bowl
Full of a steaming yellow liquid –
A mustard bath.

Gran's gone now.
Instead of mustard powder
We buy our mustard in pots
From the supermarket.
When I had chilblains,
The doctor gave me a prescription
For an ointment from the chemist.
As I rubbed it on
I thought of gran,
The white enamel bowl,
The yellow liquid,
Her feet swollen and chapped,
Her mittened hands.

John Foster

Uncle Bernard

My favourite uncle saws up his life into stories:
In a voice as deep and rich as Christmas cake
He tells of surrendering to the orchard's owner
After being sighted stealing apples
Rather than face my grandfather, that moral man;
Of running away from the job my grandfather got him,
Reporting on the local evening paper,
To repairing waggons and watching the trains
Run life-size by his hut beside the line.
His terraced house is full as an auction
Of furniture full of carving full of dust
In a road the image of Coronation Street.
Any dirty trick draws from him
A deep, intimidating growl,
Still the tough fullback who in his younger days
Was solid in defence more ways than one,
In later years he makes the best night-watchman
For miles around, who brews up with the tramps
And sees the late-night yobbos off.

Stanley Cook

Child with a Cause

My grandmother was chicken-plump.
She wore long earrings, smelled of
Pear's soap and lavender water.
She kept cream in a jug under
a blue-beaded net.

Grandfather kept us both
on a tight rein, our place
at the kitchen sink. When Gran's mind
slipped slightly out of gear
I was her memory.

Nearly always, that is. She peeled
potatoes once, put them ready
for Grandfather's tea and forgot
to light the gas. He was furious.
I saw Gran's tears.

Upstairs, in the narrow hall
I waited, scuffing the turkey-red rug.
He took his time. The flush thundered.
His shape vultured against
the door. I was raw

as carrion. 'It's not fair.
You made Gran cry.' He lunged at me.
'How dare you, child? How dare you
speak to me like that?' Picked clean
by anger I ran.

'Don't mind him,' my grandmother said.
'He likes his tea on time.' The matter
was closed. Grandfather tore into
his beef stew and mashed potatoes.
I pushed my plate away.

Moira Andrew

Three or so

Is that child in the snapshot me?
That little girl in the woollen dress
By a broken door in a tiny yard
She's shy and laughing and ready to run
And shielding her eyes from the morning sun
I've forgotten the dress, and the colour of it
I've forgotten who took the photograph
I've forgotten the little girl, three or so,
She's someone else now, to be wondered at,
With my mother's eyes and my own child's hair
And my brother's smile: but the child who's there –
The real soul of her – fled long ago
To the alley-way where she mustn't go
Through the broken door that I never forgot
    Rough men on motorbikes, not to be looked at
    Scrawny cats scratching, not to be touched
    Down to the railway-line, never to go there
    Or up to the road where the traffic rushed
    Stay close in the yard with the sun in your eyes
    Come and be still for your photograph.
I can hear now the drone of those bikes
And the loud dark voices of the men
And the howl of the tomcats on their prowl
I can hear the scream and shush of the train
And the whooshing of traffic on the road
But the summer buzz in that tiny yard
And the child who laughed with her best dress on
And the voice that told her to stand in the sun
And the click that pressed the shutter down
Have gone
As if they had never been.

Berlie Doherty

The Expendable

Rambo was never forced to take
his baby brother to the jungles of 'Nam.

I told my mother but she wouldn't listen,
said she was getting her hair done
and he'd have to tag along,
but he wasn't to play any violent games
or tread in anything nasty in the park.

I told him what it meant,
being an expendable – when we shoot,
you fall down, drop to the ground,
no messin'. Lay down and die
when we tell you, then you can play.

I'd seen them in films, the sidekicks,
the men who make up numbers,
flanking our hero as he enters town
and you know they'll be the first to go,
the first to twitch and fall to the floor
as an opening salvo finds its mark.

And now, in the park, we play tough,
no one minds being roughed a little.
The gang won't care if you do as I say,
play dead, stay down and concentrate,
we can't have a corpse picking daisies
or with fingers exploring its nose.

This branch can be your gun (not that
you'll need it for long) and I'll show you how
a spectacular death could bring our game
to life. Now mind you remember what I said,
and one more thing, you're a soldier now,
leave your teddy behind.

Brian Moses

Two Snapshots

Me and my best friend
we like to go down to the station
and get our pictures taken
in one of those photo machines
you know, where you
put the money in, then
sit on the seat inside and
draw the little curtain
waiting for the four flashes.

Me and my best friend
we like to pull funny faces
and act silly, wearing
red noses and paper hats
with hello sailor printed on them
and get our pictures taken
together like that, two
for the price of one, like,
and sometimes what comes out

nearly kills us with laughing, or
frightens the living daylights out of
me and my best friend.

. . .

I was taking a picture of the harbour
when the heavy camera slipped a little
between my hands. Fortunately, I did not
drop it. But — smile, please, say 'cheese' or
'stewed prunes' — the shutter had already clicked.

Among the snaps of friends and family groups
that was the most interesting shot of all —
a tilted horizon, a leaning lighthouse almost up-
side down in a dragonish cloud, and at the bottom
just the top of somebody's straw hat.

James Kirkup

Doing Nothing Much

I could potter for hours on a lonely beach
Picking pebbles to roll in my hand,
Wondering where will the next wave reach,
Writing my name in the sand.

Near the tumbling weir, where the hawthorn's pink,
I could sit for hours in a trance
Watching the water stream to the brink
And the white foam pound and dance.

Or high on a headland find me,
While a seagull wheels and dips,
Gazing for hours out to sea
At islands and smudges of ships.

Eric Finney

First Visit to the Seaside

I

The new day
Flooded the green bay
In a slow explosion of blue
Sky and silver sand and shimmering sea.
Boots in hand, I paddled the brilliancy
Of rippled wavelets that withdrew,
Sucking my splay grey
Feet in play.

II

It was magic – the brightness of air,
the green bay and wide arc of the sea,
with the rock-pools reflecting my stare
and a maze of wind-sculpted sand-dunes where
slum streets and the Quayside should be.

It was music – not only the sound
of the buskers outside the pub door
and the band on the pier, but the pound-
ing of waves, the loud kids all around,
and gulls screaming shrill on the shore.

It was magic and music and motion –
there were yachts sweeping smooth in the bay
and black steamers white-plumed in mid-ocean;
and ice-cream, candy-floss and commotion
as the Switchback got under way.

III

The spent day
Drained from beach and bay
Green and silver and shimmering blue.
On prom and pier, arcade and b. & b.
The looped lights dimly glowed. And I could see
Stars winking at me, glimmering through
The sky's moth-eaten grey
As if in play.

Raymond Wilson

Sea Song

Oh, I'd sail the sighing seas, my love,
Come drift on the tides with me;
For I still long for the wild waves' song
And the silver fish of the sea.

Oh I'd sail the sighing seas, my love,
Where the wild weeds gently glide;
But I'm afraid of the forest shade
Where the silent fishes hide.

Don't fear the sauntering seas, my love,
As they dance beneath the breeze;
In the moonlit foam we'll make our home
Like the silver fish of the seas.

Oh, I'll sail the rolling seas, my love,
And sigh for the cry of the wind;
But what if I weep on the ocean deep
To tread a greener land?

Oh, you'll love the roaring seas, my love!
Come ride the swell with me,
Where the breaking sky is drawn to die
With the silver fish of the sea.

If I ride the raging seas, my love,
Then will you follow me.
Or will you stay till your dying day
With the silver fish of the sea?

Oh, I must ride the wild, wild seas
And you must let me be;
Till my dying day I'll roam the spray
With the silver fish of the sea.

Judith Nicholls

Sea Talk

Inside the little harbour, on the tide
That washes stones where weedy limpets cling,
I thought I heard where sleeping rowboats ride,
The little fishes' tiny whispering.

I thought I heard, beside the wooden pier,
The starfish heave a long and salty sigh
And murmur in the mussel's shelly ear,
Its longing for a bright and wider sky.

I thought I heard the underwater shouts
Of gleeful creatures . . . moans and barks and squeals,
The dolphins thrusting long and smiling snouts
And gossiping to sleek and agile seals. . . .

The noises from the restless waves and spray
Of armoured crabs that guard their rocky spots,
The sound of white sea horses at their play
Or lobsters' prayers within their captive pots.

I wish I knew that such a thing could be —
To know the songs of moving fin and scales,
The liquid language of the living sea
And hear the gentle voices of the whales.

Max Fatchen

Marie Celeste

Only the wind sings
in the riggings,
the hull creaks a lullaby;
a sail lifts gently
like a message
pinned to a vacant sky.
The wheel turns
over bare decks,
shirts flap on a line;
only the song of the lapping waves
beats steady time . . . .

First mate,
off-duty from
the long dawn watch, begins
a letter to his wife, daydreams
of home.

The Captain's wife is late;
the child did not sleep
and breakfast has past . . .
She, too, is missing home;
sits down at last to eat,
but can't quite force
the porridge down.
She swallows hard,
slices the top from her egg.

The second mate . . .
is happy.
A four-hour sleep,
full stomach
and a quiet sea
are all he craves.
He has all three.

Shirts washed and hung, beds
made below, decks done, the boy
stitches a torn sail.

The Captain
has a good ear for a tune;

played his child to sleep
on the ship's organ.
Now, music left,
he checks his compass,
lightly tips the wheel,
hopes for a westerly.
Clear sky, a friendly sea,
fair winds for Italy.

The child now sleeps, at last,
head firmly pressed into her pillow
in a deep sea-dream.

Then why are the gulls wheeling
like vultures in the sky?
Why was the child snatched
from her sleep? What drew
the Captain's cry?

Only the wind replies
in the riggings,
and the hull creaks and sighs;
a sail spells out its message
over silent skies.
The wheel still turns
over bare decks,
shirts blow on the line;
the siren-song of lapping waves
still echoes over time.

Judith Nicholls

The Moonwuzo's Sea Song

Who is that walking on the dark sea sand?
The old Bride of the Wind

What is that staring out of the weedy pool?
The newborn Monster in its caul

What is that eerie chanting from the foam?
The mermaid's gardening song

What is that shadow floating on the water?
The Fish-King's daughter

Who bears those candles down by the Sea's curled rim?
The children going home

Cara Lockhart Smith

Beachcomber

Monday I found a boot –
Rust and salt leather.
I gave it back to the sea, to dance in.

Tuesday a spar of timber worth thirty bob.
Next winter
It will be a chair, a coffin, a bed.

Wednesday a half can of Swedish spirits.
I tilted my head.
The shore was cold with mermaids and angels.

Thursday I got nothing, seaweed,
A whale bone,
Wet feet and a loud cough.

Friday I held a seaman's skull,
Sand spilling from it
The way time is told on kirkyard stones.

Saturday a barrel of sodden oranges.
A Spanish ship
Was wrecked last month at The Kame.

Sunday, for fear of the elders,
I sit on my bum.
What's heaven? A sea chest with a thousand gold coins.

George Mackay Brown

Rainbow's End

He'd been told
there was gold
at the rainbow's end,
some El Dorado chest
undiscovered and waiting.

He was hungry for gold:
Already he'd spent it
ten times over
in toyshops, in ice-cream parlours,
the stuff dreams are made of.

Like some modern forty-niner
he'd trekked
over fields and fences
to measure out his claim
on a patch of tufted grass
and cow parsley.

Till digging down
his spade uncovered
a high - heeled shoe,
an old bike wheel
and an unexploded item
from World War II.

'Lucky to be alive'
warned the army men
who dealt with his find,
dreams of rainbow wealth
exploding behind him.

Brian Moses

Chanterelles

Beginning: a lecture enlivening
a history lesson. Of Bronze Age burials,
brooches, beakers, gold and ghosts,
secret under the barrow mounds.

Later: a birthday gift,
a metal detector. Immediately successful
with a Victorian penny, a silver spoon,
buried a good foot down.

Finding: a mound on the moor,
with a myth of a golden ghost.
Chanterelles growing there, greedy,
gold apricot, scented and sweet.

Searching: imagining gold riches,
bracelets and rings, old cups and knives.
The treasure-machine singing,
digging the black, loamy soil.

Difficult: aching, feeling rather alone
with the sky darkening. Chanterelles
shining like money. The spade
suddenly opening the old grave.

Lastly: climbing into the dark
where the thin-armed, gold-boned ghost
waited with gold-coin eyes alight.
Treasure seeking too, under the mound.

Rose Flint

BEWARE KEEPERS · AROUND ARE LITTERED RUBIES · SWORDS A

Dragonkeeping

To keep a fell creature
of green light, dark and bright,
of starry scales, the glittering breath
scarlet, the claws fierce as spite –

House it secret, in shadow cave
sleek with streaky coral, sooty jade.
Nest it warm, silver woven
webbed with treasure, spilled and stolen.
Feed it dainty, milk and honey
almonds, saffron cakes and golden money.

So content your creature
soft sleeping, folded wings . . .
but one eye wary. Beware keepers;
around his dreams are littered rubies,
swords and bones and signet rings.

Rose Flint

Talk about Caves

Talk about caves! Tell us,
tell us about them!
What's a cave, what's it like?

'My strongroom, mine,' said the Dragon,
'where I hid my gorgeous gold!'
But he lay gloating there so long,
in the end he turned to stone –
crawl down his twisting throat, you can,
for his breath's quite cold.

'My house once,'
whispered the Caveman's ghost.
'O it was good
wrapped in fur by the fire to hear
the roaring beasts in the wood
and sleep sound in earth's arms!
(If you find my old knife there,
you can keep it.)'

'My bolt-hole from the beginning,'
Night said,
'where I've stayed
safe from my enemy Day.
I watch through a crack the sun
beating away at the door –
"Open up!" he shouts.
He'll never get in!'

'My home, always,' said Water.
'I wash my hands here,
and slow as I like I make
new beds to lie on
in secret rooms
with pillows and curtains
and lovely ornaments,
pillars and plumes,
statues and thrones –
what colours the dark hides!
I shape earth's bones.'

'Don't disturb me,' the Bat said.
'This is where I hang my weary head.'

Libby Houston

And They Shall Enter Caves

Pursued by wolves
they followed the waterbrook
into the rock

down tunnels
whose smooth walls glowed
with cold green light

and on the cavern floor
found fragments
of heroes' swords
in dragon fossils

Roger Lang

The Dare

Go on, I dare you,
come on down!

Was it *me* they called?
Pretend you haven't heard,
a voice commanded in my mind.
Walk past, walk fast
and don't look down,
don't look behind.

Come on, it's easy!

The banks were steep,
the water low
and flanked with oozing brown.
Easy? Walk fast
but don't look down.
Walk straight, walk on,
even risk their jeers
and run . . .

Never go near those dykes,
my mother said.
No need to tell me.
I'd seen stones sucked in
and covered without trace,
gulls slide to bobbing safety,
grasses drown as water rose.
No need to tell me
to avoid the place.

She ca-a-a-n't, she ca-a-a-n't!
Cowardy, cowardy custard!

There's no such word as 'can't',
my father said.
I slowed my pace.
The voices stopped,
waited as I wavered, grasping breath.
My mother's wrath? My father's scorn?
A watery death?

I hesitated then turned back,
forced myself to see the mud below.
After all, it was a dare . . .
There was no choice;
I had to go.

Judith Nicholls

The Weasel

It should have been a moment
Of high drama,
The lithe, cigar-slim body,
The slight withdrawal and swift spring,
The soft explosion of black feathers . . .
But when I glided to a halt
And leaned from the car window
The blackbird had lost the duel
Of dignity,
And was being dragged,
Claws drooping palely
Upthrust on straw legs,
By the tiny killer,
Rump high, teeth gum-deep in feathers
Just above the two
Baby finger-nail eyelids.

The click of the ignition key
Startled the weasel
And through the verge grass he disappeared,
While the incongruous corpse
Stiffened in the dust.
But patience.
This plump prize whose blood
Is still, but still warm,
Is being watched,
And three anxious minutes pass
Before the grass
Parts,
The triangular, fawn head appears,
Looks up and down the lane,
Ignores the car,
And leaps to the carcass,
Drags it up the small cliff of the bank
And slides through the hedge backwards –
The whole action so neat,
Cool and efficient,
The work of a professional.

Gregory Harrison

THE·WEASEL·

The Fox and the Egg

I held a hatching egg against my ear
And listened to the cheep of bird inside.
I heard the chip-chip of its tiny beak
The urgent hammer as it struck for air.
I put the egg, warm from my nestling hand
Back in its bed of straw. Tonight
It would thrust out bony wings
And twigs of legs, and crack its shell
And peer blindly into life.

But then, as we all slept,
The fox leapt the six-foot wire fence
Crept inside the shadows of the barns
And took the new-hatched chicks, whole,
And bore them home to feed his new-born cubs.

I tried to forgive the fox.
I thought of his den hidden in dark earth,
I thought of his red cubs warm by his vixen's side,
I saw their hungry puppy-mouths wet and wide.

The fox streaks through the forest
The fox slinks through the hedgerows
The fox flees from the hounds
The fox spills his blood.

The man who chased the hounds had an evil heart
Killed for sport.
The hounds who chased the fox had animal hearts
Killed for blood.
The fox who stole the chicks had a cunning heart
Killed for food.

And I
Still hear the cheep of the cracking egg
Still feel the warmth of it in my palm.

Berlie Doherty

114

War

trees cows walls
blown out like paper
like so much litter
balled up
in
the
wind

and after it the clear flat surface knowing no one

John Sewell

War Scene

The mist-shrouded trees
Splintered and mutilated.
The men on their knees
Broken and bewildered.
The Generals
Grim, but contented.

Ray Mather

Burma Star

When I was little, knee-high
to an ice-cream van, I wore
a cowboy suit of blazing
red with yellow buckskin fringes –
a regular Roy Rogers.

My father, smiling,
pinned his Burma Star,
his other bronze medallions,
across my pigeon chest.
Like sheriff's stars I wore them.

I lost the medals, spinning,
falling in a graceful arc,
cut down by phantom gunmen from
a wild, imaginary West.

My father shrugged
and rubbed my hair
and for an instant, in his eyes,
I saw real soldiers coughing
blood-red death within the dark
and steaming jungles
of his past.

Phil Carradice

117

Just Another War

On her sideboard
Nan has a picture
Of a young man
In a soldier's uniform
Smiling proudly.

'That's my brother,
Your Uncle Reg,'
She says,
Her voice tinged
With sadness.

'He was killed
In Korea.
He was only nineteen.'

'Where's Korea?' I say.
'What were they fighting for?'

'Somewhere in Asia,'
She says.
'I don't know.
It was just another war.'

John Foster

Dis Fighting

No more fighting please, why can't we stop dis fighting,
dis fighting hurting me, why don't we start uniting,
dem fighting in Angola, dem fighting in Manchester,
dem fighting in Jamaica, and dem fighting in Leicester,
well i might be black, my people were once slaves,
but time goes on, and love comes in,
so now we must behave,
it could be that you're white, and i live in your land,
no reason to make war, dis hard fe understand,
skinheads stop dis fighting,
rude boys stop dis fighting,
dreadlocks stop dis fighting,
we must start uniting,
our children should be happy and they should live as one,
we have to live together so let a love grow strong,
let us think about each other, there's no need to compete,
if two loves love each other then one love is complete,
no more fighting please, we have to stop dis fighting,
dis fighting hurting me, time fe start uniting,
dis fighting have no meaning, dis fighting is not fair,
dis fighting makes a profit for people who don't care,
no more fighting please, we have to stop dis fighting,
dis fighting hurting me, the heathen love dis fighting.

Benjamin Zephaniah

The Lord's Lament

A man said unto the Lord
'Lord, may I build a house in your garden?'
'Yes' answered the Lord, 'You may'
and all was well
but not for long
for man was in fear of the wild animals
that walked freely in the garden
and thus asked the Lord
'Lord, may I build a castle to protect myself?'
'Yes' answered the Lord, 'If you are so afraid'

and once more all was well
for a while
but then
'Lord, may I build a factory so that I may work
and create?'
'Yes' answered the Lord, 'If you so desire'
and for a time man was again content
but the factory created curiosity
so much so
that man asked the Lord another question
'Lord, may I dig up the Earth and experiment
so that I may better my life?'
'Yes, if you think you have need' replied the Lord

and so man did
but as well as growing in curiosity
he also grew in greed
he was no longer content with the garden
and asked of the Lord
'Lord, may I explore space
and the seas that surround the garden?'
'If you so wish' answered the Lord
and man did wish
until soon, he believed he could do anything he wanted
and began to do things
without first asking the Lord's permission
man also began to wonder about himself
and set himself impossible tasks
and asked himself unanswerable questions

Then one day
man asked the Lord one more question
'Lord, if I destroy myself
will you make me again?'
this time the Lord did not answer
but his tears rained down upon the Earth
for forty days and forty nights
the second time he had cried for man
and perhaps the last.

John Walsh

121

Buddhist Parable

Because of a life of evil deeds,
a certain man now finds himself in hell.

Yama, king of those infernal regions,
questions him severely, saying:
'In all your wicked life, did you ever encounter
any messengers from heaven?'

The man at once replies: 'No, my lord,
I never met such ones in all my life.'

Then Yama asks him, with piercing gaze:
'Did you ever meet people, old and infirm,
bent with age, and walking with a stick?'

The man replies, without hesitation:
'Oh, yes, my lord, I often saw such persons.
You can find them everywhere on earth:
no one takes any notice of them there.'

Then Yama tells the careless fool:
'The reason you are now suffering
the torments and the agonies of hell
is that you passed those people by
without a word, without a thought,
and did not recognize
in that old man or that old woman
a heavenly messenger sent to warn you
of the evil of your ways,
and that you must change them
before you, too, grow old.'

James Kirkup

The Boy

The boy,
Trying to catch Time in his arms,
Stumbles,
To find he is a man.

Ray Mather

At the World's Imagined Corners

Having only just arrived, late the night before,
I hesitate the next day at the front door
Prior to going up the road to find a post box.
I peer into the pale morning light
At an unknown street in a strange country.
If I set out, will I find my way back all right?
Or if I keep on walking could I be lost for ever, disappear?
Is it still possible that old temptation, that old fear
Of falling off the edge of the world?

John Cotton

Index of first lines

Acknowledgements

The following poems are appearing for the first time in this collection and are printed by permission of the author unless otherwise stated.

Moira Andrew: 'Full House' and 'Nature Study' both © 1987 Moira Andrew. James Berry: 'Benediction' and 'Boxer-man in-a skippin workout', both © 1987 James Berry. Valerie Bloom: 'Duppy Story Time', © 1987 Valerie Bloom. Dave Calder: 'Grandfather Gavin', © 1987 Dave Calder. Stanley Cook: 'Acrobats', 'The Watertower' and 'Uncle Bernard', all © 1987 Stanley Cook. John Cotton: 'A Snowing Globe', 'In the Woods' and 'At the world's imagined corners', all © 1987 John Cotton. Peter Dixon: 'Carps', © 1987 Peter Dixon. Berlie Doherty: 'Three or so' and 'The Fox and the Egg', both © 1987 Berlie Doherty. Max Fatchen: 'Sea Talk', © 1987 Max Fatchen. Reprinted by permission of John Johnson (Authors' Agent) Ltd. Eric Finney: 'Doing Nothing Much', © 1987 Eric Finney. Robert Fisher: 'To Find a Poem', © 1987 Robert Fisher. Rose Flint: 'Chanterelles' and 'Dragonkeeping', both © 1987 Rose Flint. Pamela Gillilan: 'The Invalid', © 1987 Pamela Gillilan. Julie Holder: 'Eyes' and 'Silent Shout', both © 1987 Julie Holder. Geoffrey Holloway: 'Dawnsight', 'Walking Out' and 'Blue Tits', all © 1987 Geoffrey Holloway. Jean Kenward: 'Snow', © 1987 Jean Kenward. James Kirkup: 'Two Snapshots' and 'Buddhist Parable', both © 1987 James Kirkup. John Kitching: 'The Touch of Sense', 'Black' and 'Request', all © 1987 John Kitching. Roger Lang: 'And They Shall Enter Caves', © 1987 Roger Lang. Ian McMillan: 'Goodnight Stephen', © 1987 Ian McMillan. Ray Mather: 'Ordering Words', 'Cross-country', 'War Scene' and 'The Boy', all © 1987 Ray Mather. John Mole: 'The Whole Duck', © 1987 John Mole. Brian Moses: 'Classroom Globe', 'The Expendable' and 'Rainbow's End', all © 1987 Brian Moses. Judith Nicholls: 'Notes towards a poem', 'Marie Celeste' and 'The Dare', all © 1987 Judith Nicholls. Grace Nichols: 'Break/Dance', © 1987 Grace Nichols. Joan Poulson: 'Moor-walk', 'Our Farm, January' 'Sea Spring' and 'Bird-cloud', all © 1987 Joan Poulson. John Rice: 'Let you body go', © 1987 John Rice. Albert Rowe: 'The Starlings', © 1987 Albert Rowe. Vernon Scannell: 'Waste-paper Words', © 1987 Vernon Scannell. John Sewell: 'Seven Days of Spring' and 'War', both © 1987 John Sewell. John Foster: 'Invaders', 'Gran' and 'Just Another War', all © 1987 John Foster.